TI(

GTAW need to know for beginners & the DIY

home shop

2nd Edition

Spencer Gould

Prologue:

This book is aimed at people new to TIG welding who are mostly learning for their own uses and at those who have been away from it for quite some time and are looking to get back into it. Learning TIG may seem quite intimidating but if you're not getting the results your looking for in your welds this book can help you get into this fabrication method and guide you in the direction of mastery. The book is setup in a need to know format and a first pass read takes about an hour or two in total. While you can hunt down and gather this info for free from other sources across the vast ocean known as the Internet as I did at first your looking at a 10 - 100X investment of your time with a good amount of confusion and the potential to head down the wrong path. So, in the end what's your time worth? Success leaves clues. This book can be your map for the journey.

If you're going to tech school / college to become a professional welder, this book may be of some assistance as an extra boost, however most of this material will be covered in your course and lab work just strung out over months.

If you're already a professional TIG welder with Lincoln Electric or Miller tattoos on your biceps, flaming skull welding helmet to match, and a "TIG Life" bumper sticker on your Hot Rod I admire your skills and dedication but this book would be too elementary for someone of your skill level.

I went through the white-collar education and career path of aerospace / mechanical engineering but have found a great amount of value and competency learning blue collar skills like welding, machining, woodworking and composite layups. I have found time and time again that the most valuable co-workers have outside-of-work, hands-on skills with auto / aero / marine or civil endeavors. The Pure Theory / no practical application people quickly show their lack of value with in a technical organization.

Why learn how to TIG weld? If you're reading this, you likely have some level of interest in learning how or have a need that TIG is the best option for. TIG can lead to a strong primary career path or something you can do as a side job. As mentioned in the introduction, it can be a highly valuable skill to have as a mechanical engineer. And from an American pride/looking out for the country perspective, the USA general population has slowly been converted from producers and creators to general consumers, where everything is an "app" or an easy option where we buy premade products with no idea of what's in them or how to make them. For our own national and economic security there needs to be a shift in our culture to where citizens know how to create not just buy. Other countries are also falling into this trap, so the more hands-on skills your citizens have the more of an asset it is for your country. Even with advanced composite materials and 3D metal printing, there is still a very strong need to use arc welding methods to create products, vehicles and structures.

Since the initial release of this book I am proud to say that its helped people start on their own welding journey and beyond the USA, the book has sold across the globe, with particular interest in England and Australia.

The purchase of this book in either paperback or Kindle formats is of a static copy. I have a companion web page setup to cover some of the latest projects, resources, links and to Amazon and other online pages where you can purchase TIG welding machines and equipment, all filtered and focused for the beginner and those looking to setup a small shop (for home and business). I've researched, shopped and purchased a good amount of manufacturing equipment not only for my own needs but also for many of the companies I've worked for so this can be of great assistance to you the new or returning TIG welder. www.gouldaero.com/weld can act, for you an additional resource for years into the future.

Table of Contents

The Big Question:

Do I need to go to a professional school to learn how to TIG weld? If it's to start out or for your own uses, then my opinion is, no, you don't. The most applicable analogy I can think of is this: do you need to go to a professional race car driving school to learn how to operate a manual transmission? The answer is, of course not; millions of people across the globe have learned to drive a manual with the most informal of education, however this is not a substitute to do this in a vacuum. Just like a manual transmission car, TIG Welding is an exercise in hand/eye/foot coordination. For most people new to the world of welding, the best first hands-on opportunity to see or try the machines is large aircraft, car and motorcycle shows. Being an airplane guy, I've been to Oshkosh and Sun 'n Fun multiple times, and Lincoln Electric usually has a display where you can get some hands-on experience with their product line. If you approach this endeavor 1) with a focus on safety, and 2) you keep learning and practicing to master high quality welds, you will do well.

Note, there are some good YouTube welding channels out there but you would need to binge watch for weeks, like I had to, to put together the comprehensive information included in this book. Many of these YouTube welding channels have very talented welders, but if you're getting started it can be QUITE frustrating trying to learn it yourself with just a video or two. Also, people love to consume free internet and YouTube content, however this information is very rarely acted upon. On the other hand, paid for content (books, courses and workshops) typically have a higher percentage of people taking action and actually doing something with the information.

This book can be an excellent place to start or catch up. If you get into the groove of things and are thinking of taking on welding as a career, it could be a good way to go. At this point many of the higher-end welding jobs will require, at the minimum, certifications. You can practice for the test and go straight for the cert or go to a tech school, at this point. Try to be adaptable throughout the process and not fall into bad habits. Getting the basics down on your own time with your own machine can give you a big leg up over people starting from zero. Plus at the end of the day, you will have a TIG machine.

Pro's and Con's

TIG benefits over other welding methods:

Compared to MIG, Stick, and Oxy Acetylene, TIG offers the cleanest/highest strength-to-weight welds with the lowest heat distortion. From a "do it at home" perspective, the fire risk is the lowest compared to other methods. Sparks are almost non-existent with correct gas flow and the "ball of heat" is about the size of a golf ball for most car and aircraft fabrication projects, compared to showers of sparks or long flames that happen with the other welding processes. TIG also offers the most control of the weld, allowing you to vary the heat, filler feed rate, and weld profile shape. TIG is a clean welding process where you can start and stop wherever and whenever you want. This compares to Stick and Flux Core MIG that form a nasty scale that protects the solidifying weld from the atmosphere. With those welding processes, you need to remove the scale completely before going anywhere near that area again.

Disadvantages of TIG over other welding methods:

While TIG may be seen as a viable production method on short/highly critical welds like engine mounts, roll cages, bicycles, etc, other methods surpass it in longer/less critical welds like ship, general auto and civil construction. TIG is also the most difficult hand welding process to master as you must:

- Carefully control, at all times, the distance from the electrode to the work piece. Make contact with the electrode and you're done for a while. The electrode will need to be re-ground to the correct profile.
- Load filler into the weld puddle with out making contact with the electrode
- Control amperage typically through a foot pedal.

Pro's of learning the skill and buying equipment: To sum it up, it's about confidence, capability and convenience. Like learning how to run and own machining, equipment you won't know how you ever got through life without it. If something metal breaks, welding it back together is now an option at your disposal. Need some custom-made thing you can't buy or that is ultra-expensive or hard to get? You now have the means to make it happen. You can use the best fabrication method for the job. Paid-for welding services can be expensive; most shops have shop minimums and do not cost effectively fit into the "odd job", "one or two welded component" models. Most TIG welding shops want four, five or six figure jobs for the mainstay of their work to make it worth their time as a business. Before learning TIG and having the machine I made quite a few parts out of other methods like carbon fiber or riveted aluminum that would have been ten times easier to make by welding it from 4130N Chromalloy.

For example, here is a carbon fiber rudder pedal and idler arm I made before learning how to TIG weld. It would have been ridiculously easy, comparatively, to make it out of 4130 Chromalloy, like in the majority of light aircraft.

An example of conventional welded 4130N rudder pedals

Con's of learning the skill and buying equipment:

For some, learning new things can be scary. Others lack confidence in their own fabrication abilities. If you missed out on Legos as a kid, this maybe you. On the equipment side, beyond the basic machine cost, helmet, and required accessories that typically add up to two to three grand, the big issue for many is where to keep and use their TIG machine. If you own your own home and don't have an overzealous HOA, you should be in good shape, however, if you're in an apartment or living with people (spouse, parents, roommates etc.) dead-set against "that kind of thing", you could be in for a big challenge.

Note: Laser, spot and stir friction welding are excluded from the above, as this field is largely dominated by automated CNC equipment.

Shop Safety

Since the initial publication of this book one of my fellow EAA members bought the book, breezed over the safety section, but did read through the rest. He took action, bought equipment and began to use it. He used old school oxy acetylene goggles, a t-shirt and shorts and soon got the mother of all sunburns that lasted for months. Arc welding safety is fairly straightforward once you know the rules. Think with the mind set of treating it like a loaded gun or a running chain saw, and you will be going in the right direction.

ELECTRIC ARC WELDING IN PROGRESS

TIG welding is a form of Arc welding where an electric circuit is created that is intended to arc between the tungsten electrode and the work piece in the area you want the weld. Dangers can come up when your body is the path of least resistance for the electric circuit, particularly with exposed skin. There is way more than enough current in typical gauge thicknesses to kill or cause injury. The first safety item is to have your grounding clamp on your work piece, or if not feasible, on the metal worktable on the opposite side of the TIG torch from you. Your working area should also be dry and free from standing water.

The other issue with the arc is you being exposed to dangerous levels of UV light that can cause sever sun burns and with long enough exposure, skin cancer. Make sure to inform anyone that could be coming in your shop while you're welding to not look at the Arc with the naked eye or get too close. Most professional fabrication shops have blocking shades / curtains up to protect bystanders from the weld processing. While many of the "cool dudes" just have a helmet on, the exposed skin in contact with the work piece or welding table makes for an easy conduction path, not to mention, skin that looks like leather later in life requiring, multiple MOHS surgeries.

A quick note: speaking of voltage, always check that the supplied AC frequency in your own country is a match for what the machine is set up to handle, not just the volts and amps. In the US we are on 60hz, while much of the rest of the world is on 50hz. Always make sure that your electrical supply is up to your local building codes. These machines can draw some serious power from your shop's electrical system, so make sure its up for the challenge.

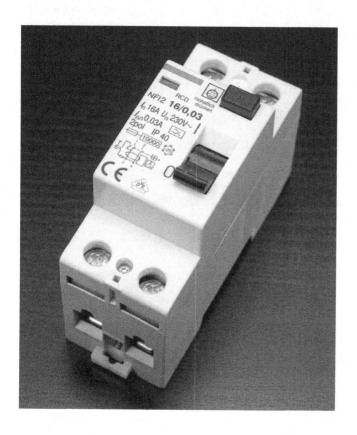

Some chemicals SHOULD NEVER BE USED TO CLEAN METAL TO BE WELDED! This includes brake cleaner or anything with even the smallest amounts of chlorine as this will vaporize, producing toxic fumes.

Stick to things like Dawn dish soap to remove grease and isopropyl alcohol for the final pass over the to be welded areas.

Even with proper metal preparation TIG welding produces fumes as various coatings / oxides vaporize. Fume extraction and or protection methods should be utilized at all times. This is a delicate balance, as the argon shielding gas can easily be blown off the weld with a strong breeze or fan pointed at you. For heavy amounts and long duration welding you should have a fan in a window setup that pulls the air out of your shop to the outside. Many of the welding OEM's offer professional fume extractors where a cone shaped nozzle is placed near the work piece to pull the fumes away. I ended up using an exterior window vent plus a box fan with a carbon filter duct-taped to the discharge side of the fan. With this setup, you can visually see the fume smoke being pulled in the fan and stopped by the carbon filter. For tubular truss welding, you can usually cap or plug the free ends with aluminum foil or machined aluminum plugs that capture the fumes inside the tube. Advanced TIG setups have plugs on the open ends of a tube cluster with a side supply of argon that purges the oxygen from inside the tube for an even higher weld. This is on the elite side of thing but there are millions of TIG welded tube structures done with out this setup.

Welded components will stay hot for several minutes after finishing the weld. The more welding done at one time on a part or the thicker the part, the longer and hotter the part will stay. Aluminum has higher heat conduction vs. steel, so it stays hot even longer.

Optical metal temp sensors may read or read cold a welded metal part, particularly if its aluminum, so I would not trust this route. Exercise caution when handling these parts. Treat it like a hot stove; test before you grab.

TIG welding requires shielding gas, typically argon. This displaces oxygen away from the weld. However, an argon leak in a work shop or car can displace the oxygen in the room making it difficult or impossible to breathe in a small area. The argon cylinder is highly pressurized (up to 2,000 PSI) and requires careful transportation and storage. When at your shop the cylinder to be chained or strapped to your welding machine, cart (if provisions exist), or a column or stud-mounted hard points against a wall. Unless you have a pro welding rig or truck, the safest way for you (though, maybe not other people) is to transport the cylinder going sideways and never with the pointy or blunt end pointing towards you or any other passengers in the car. Pointy way behind and towards you can impale you in a wreck: blunt way towards you would be bad if the neck of the cylinder ever broke off and it became a rocket.

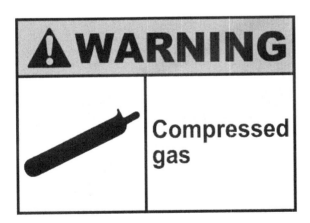

Excluding intentional jigs and tooling fixtures that could be made out of wood, you need to clear near-by flammables, including wood, paper and card board, away from the work area. Gas, oil and solvents should be kept far, far away from the work area. For general fire safety, don't do a welding job, or anything else that involves fire or sparks, in your shop and leave for a long trip right away. You usually want to be around for a few hours after to make sure nothing lit up. Keep a charged, medium-size fire extinguisher next to your setup incase anything lights up.

Noise: DC steel welding is usually a fairly quiet process and does not need hearing protection, however, AC welding for aluminum can sound like a swarm of VERY LOUD angry bee's. Foam Plug ear plugs should be worn for AC welding. The process of grinding electrodes and prepping metal can be quite loud, so ear protection should be worn for this, as well.

Tungsten Electrodes: use eye protection, respirator and hearing protection when sharpening tungsten electrodes. Tungsten is a heavy metal and its dust should not be inhaled. A Nosh rated respirator will do the job and they are low cost and commonly available at the box home improvement stores and on line.

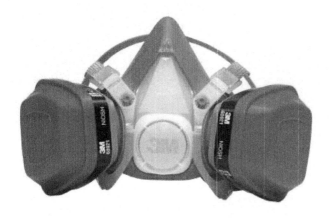

TIG tungsten comes in several different alloys. One of the older alloys is thorated tungsten "red band". The alloying element thorium is radio-active and can be of concern, however, so is non-stop cell phone next-to-head usage. Thanks to improvements in technology other non-radio-active element additives are available, like E3 "purple band" and lanthanated "blue band", both work great on most steel and aluminum.

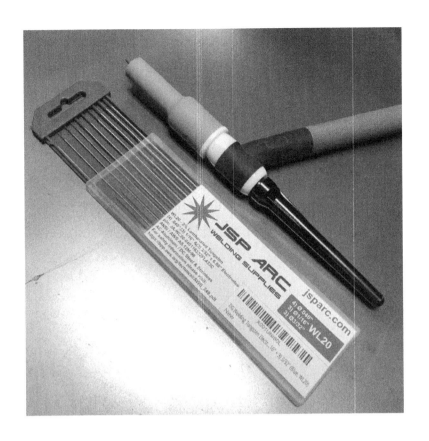

Stick welding setup caution.

Many TIG machines also have a Stick welding function. Unlike the typical TIG setup where you have to press a foot pedal or button on the torch to get things going, a Stick setup is live when the machine is on. Make sure to only insert the Stick electrode when ready to weld and if the machine is on and you are not actively welding the stow the Stick holder with the electrode out of it away from the workpiece.

The ends of Stick electrode holders are usually covered in nonconductive plastic so as long as the electrode is out and the holder is not clamped to anything metal you should be good.

Types of Machines

For more detailed information on some of the latest welding machines including where to buy them please visit the companion web page: www.gouldaero.com/weld

Transformer Machines: This is the original / older technology. From an electrical engineering perspective, these machines are on the simple, more robust side. However, almost all transformer machines need 208/240 50A power. For a home shop, this may require an electrician to come out and wire you up a 50A outlet. They are HEAVY, my friends Lincoln Precision 225 TIG used for my early welding weighs over 200 pounds. This makes it a challenge to move and store. That coupled with the power limitations, also puts limits on portability and taking the machine to remote locations. These can be overcome with more equipment and funds. (Lincoln Precision 225 Transformer Machine)

Many of the transformer machines do not have a pulsar function on
them. Accessory pulsars can be added at additional cost and size.
The reason you want a pulsar function is to control the heat going
into thin wall materials allowing you more margin before you blow
a hole through the work piece. In layman's terms, you will have to
sit in one spot for a longer time with a pulsar function before a
blow out will happen.

Inverter Machines: This is the newer technology. It has been refined over the last decade and is the type of machine I own. Many inverter machines can run on normal 110V wall outlets, as well as 50-amp services. They typically weigh between 30 to 60 pounds; are easy to store, transport, and use in remote areas. They can also be coupled with the more common 110V generators. Many inverter machines have built-in pulse and advanced AC functions.

Dedicated TIG Machines: On a dedicated TIG only machine you will usually get a foot pedal for more control over the process and some offer A/C functions so you can weld aluminum. These machines usually have the capability to process stick welding, as well. (Lincoln 200 Square Wave Inverter Machine)

Multi Process Machines: These machines are primarily MIG (wire spool or "wire feeder") machines, where TIG is an add-on function. Typically, their TIG capabilities are limited. Many of these machines have what is called "scratch start TIG", where the electrode is scratched across the work piece and lifted up. With the electrode to work piece gap as a critical item, this can be a difficult trick to pull off. Additionally, many multi-process machines will not accept a foot pedal or you will have to open up the case and modify it as an add-on kit. The other draw back when it comes to TIG is many of these multi process machines only do DC welding meaning no A/C and there by no Aluminum welding. If you are tempted by a multi-process machine and have a strong interest in TIG, carefully look over the features to see if the above mentioned applies to the exact machine you're looking at and if this will be a challenge for what you want out of the machine. Note: multi-process and dedicated MIG machines also come in inverter technology, allowing for operation off 110V and 208 / 240V. (Lincoln 210MP multi-process inverter machine)

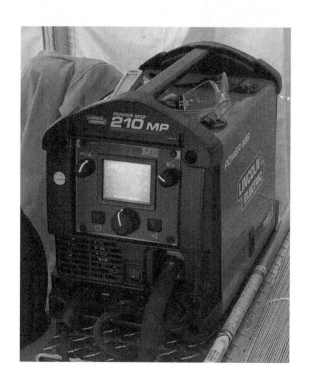

Where to Spend the Money and Where to Pinch Pennies

For more detailed information on some of the latest welding accessories, including where to buy them please visit the companion web page: www.gouldaero.com/weld

Spend the Money on:

Going with a used TIG machine may be tempting, however unless you have an experienced TIG welder friend with your best interest in mind to check out the machine with you, you could be paying for someone's high-cost, broken lawn ornament.

When it comes to buying the main machine, I recommend buying new and direct from the original equipment manufacturer (OEM) or from a reputable brick and mortar or online distributor like Amazon that typically ships the machine direct from the OEM. Shop around and learn your products. Welder machine brands can be as heated of a topic as Ford vs. Chevy, with a lot of "other brand" misinformation. With that aside focus on brands and models where the company's prime focus is making welding equipment, not just another private label to fill up a catalog. When you do find a brand and model that you like, you can look for occasional sales if you get on the OEM's or machine distributor's mailing. Major aircraft and car show's typically will have show specials direct from the OEM or a major distributor at the show.

Holidays and spring can be prime time for sales on these machines. If you have a welding supply shop near by like Praxair, you could ask to be put on a notify list if there is a particular machine you like goes on sale, as well.

Another place where you should buy the best is with your welding helmet. While there may be some new $50 helmet options out there, "ask your self are your eyes worth it"? Lincoln and Miller, along with other vendors, have a good range of safe helmets that are specialized for just welding protection equipment. While some of the lower-cost helmets may have good enough safety and work fine on MIG and Stick, the optical clarity and performance required on TIG is much higher due to the precision with which the TIG electrode and filler rod must be held to the surface, particularly with low amp/thin wall work, where MIG and Stick go in to make full contact with the work part.

TIG gloves are another thing not to skimp on. Typical MIG or generic work gloves are way too thick for good TIG work. This can have a direct effect on the quality of your welds. A .035" Ø filler rod can be difficult to feel and control with heavy gloves. A simple change in gloves can make the difference.

Other stuff you're going to need:

You need a way to cut and surface finish the metal parts you intend on making. Here are some of the cost-effective options that save a bunch of time on processing:

Cold cut chop box saws for cutting tubes, channel, and rod. I have an Evolution Rage2 Multi Purpose Cutting Chop Saw with a 14" blade. This saw has a slow rpm/carbide tooth blade that makes milling machine precise cuts with almost no sparks and low noise. Conventional fiber mesh metal cutting saws can be quite messy and throw flammable sparks long distances in your shop. While there are not too many cold cut chop box saw make and models out on the market, they are usually available on the internet or through specialty order and they're worth checking out. Cold Cut saws work equally well on steel and aluminum. Thin wall tubing processes quicker, solid rod will take some time to process, but if you let the saw do the work it can make it through.

Cold cut circular saws are great for cutting large sheets, where the throat capacity of other methods is a limitation. I have the Evolution Rage B 7-1/4" Multi-Purpose Cutting Circular Saw. Using fences to guide the saw works well and gives an improved precision on your cut.

Hand-held portable band saws, or "port-a-bands," are a good mobile option when cutting stock. I have had two Harbor Freight portable band saws and have gotten more than my money's worth out of both. The latest HFT part number for there portable band saw is #62800. Milwaukee and DeWalt also offer portable band saws. A 44" X .5 bi-metal blade is a common size and can cut through steel as well as aluminum.

Coping jig is another good item to have if you're doing round tubular truss/frame work, like engine mounts, roll cages, or bicycle frames. This allows you to quickly get to the correct cut for intersecting tube clusters.

Nice to haves:

Flat Pattern Coping: if you know and have a CAD system you can CAD model your tubular intersections and create an un wrapped flat pattern that you can usually print at 1:1 scale from an 8.5 X 11 printer, cut out the pattern, wrap around the tube, sharpie on the shape, and use your cutting method of choice to get close likely with a Dremel and sanding drum or grinder to dial in the final shape. Free CAD is an open source CAD program that has good YouTube training videos and cost noting. I personally use a full commercial seat of Rhino 3D V5. With a price point under a grand, it is packed with features that are found on CAD systems costing 5 - 20 times more.

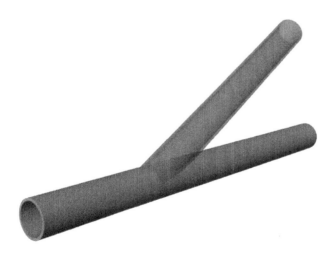

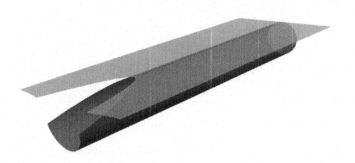

Milling Machine / Lathe Coping: With the proper setup you can mill in the coping angle. A simple example would be a 3/4" tubing joint with a 90° angle. You can clamp the tube and collet up a 3/4" end mill that will cut the intersection for you in the tube. For other angles, particularly sharp ones, you can clamp the tube up to the X/Y apron on a lathe and chuck the end mill in the lathe chuck. Some dedicated lathes may take some ingenuity to get the correct setup. Three-in-one machines that are a combo mill/lathe/drill. The lathe cutters that are typically attached to the milling X/Y, table so this makes a vice setup way easier as "T" slots are already on the machine and can be used with the lathe function. Note: 4130 tubing is easily processed with high speed steel "HSS" end mills, Titanium Nitride (TiN) coatings do aid with cutting/chip evacuation. I have owned a Smithy Granite GN1340 Max (three in one lathe/mill/drill) for around a decade now and have easily gotten my money's worth out of this machine.

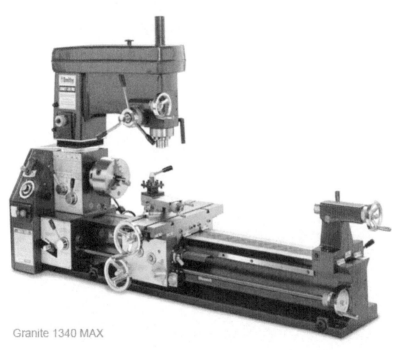

Granite 1340 MAX

Made to order tubing kits: There are some services out there that will cut the notch patterns into the end of the tubing for you. Others can provide laser or water jet cutting services. In the modern era most of these companies use CAD software between you, the customer and them, however, some may still work with drawings and paper templates. .stp "step" and .igs "iges" are the two main universal file formats to send things from your CAD system to what ever system they are using.

Where to save:

For your first at home welding setup with an inverter machine, I find a dedicated welding cart on wheels a necessity. This does not mean you need a gold-plated welding cart. I found the Harbor Freight Welding Cabinet (#61705) to be an excellent all-around cart that holds almost all my welding stuff in one place. If you do end up with a transformer machine many of these either include wheels and a base for the argon cylinder or a simple add-on. Caution should be exercised if putting a transformer machine up high on a cart, like the above-mentioned HFT model, due to much higher weight and top-heaviness compared to an inverter machine.

Harbor Freight and Tractor Supply are good for some of the welding accessories like 45°/45°/90° triangle magnets, F-Clamps, MIG pliers and welding arm sleeves, as well as some other welding supplies.

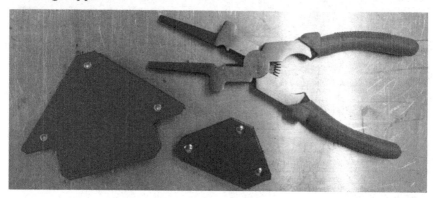

Electric or air-driven rotary tool: while a Dremel is on the spendy side of things for what it is, there are competitors that are at a lower price point. With some good sanding / grinding bits, a rotary tool is the fastest and most practical way to prep tubing for welding. Check Amazon and e-bay for rotary tool attachments like sanding drums. You can typically find large bulk packs (twenty, fifty, one hundred…) of a given attachment for the cost of only a few (six) at a box store.

Another very handy metal prep power tool to have is a 4-1/2"
angle grinder with a sanding flap disk. This works well for flat
metal, however, caution should be exercised with low grit paper
and thin wall tubing as it can plow through it.

Proper Material Prep and Cleaning

This is the very first thing that you must have correct and can be a common area where new welders have issues with the quality of their welds. Prep comes in 5 major steps:

1) You need a good part-to-part fit. The TIG process can bridge across a .050" linear gap, but it will not bridge a one-inch gap. It can be tedious work but will be one more thing going right for you when you turn on the machine.

2) Sometimes material comes with a bunch of "gunk" on it. A fresh razor blade can help take off the bulk quickly.

3) Dawn dish soap and water can help lift up remaining oily residue. Any dish soap residue left behind will not pose a safety issue once heated up as compared to harsh cleaning chemicals that should never be used in this application.

4) Clean off oils and dirt from the metal with paper towels and isopropyl alcohol for a solvent. If you grind over oil and dirt it will just spread it around and not necessarily take it away.

5) Mechanical prep:

     i) For most general, in-the-open areas, 150 - 300 grit sand paper wrapped around a foam sanding sponge works quite well.

ii) Dual action "DA" or random orbit sanders can also work in open areas. Again use 150-300 grit.

iii) Use a rotary tool with a sanding drum or grinder attachment to sand up to 1" min away on exterior surfaces and do the best you can on the inside of items like tubes for areas you are going to weld on. rotary tools will easily take away surface rust on steel. Grinder usage on a rotary tool for aluminum is generally avoided as aluminum galls and loads up the grinder wheel. Heavily loaded grinding wheels can explode causing severe injury or death. Aluminum on a bench grinder is a major shop no-no as it can make the grinder disk explode and punch through almost anything in its path including you. For aluminum prep it's best to use sanding drums, belts and burrs.

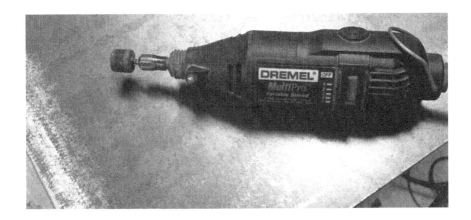

v) If you have it, a lathe works quite well on new stock round tubing. Simply chuck the tubing in the lathe jaws and use sand paper. You can get a very clean helical/honing pattern on the metal. Be sure to have a loose grip on the sand paper so that if something catches you can let go quickly.

6) On thick metal you will need to bevel the edges to allow proper weld penetration. The American Welding Society (aws.org) has a good amount of material on bevel prep.

7) Isopropyl alcohol wipes should be used to clean up any remaining dust left behind by mechanical prepping.

Remember when welding it locally liquefies the parent work piece and any contaminates. Things like rust, dirt and paint getting into the weld reduces the quality, makes it far more difficult to weld, makes for localized weak spots, and creates cosmetic issues, as well. Having a clean surface can be the difference between a good weld and a booger weld.

Hand support/proper body position: One of the biggest challenges in TIG welding is grounding out the electrode against the work piece. Even if you think you think you have the steadiest hands in the world, try holding an ink pen a 1/8" off paper with out your hands or wrist resting against anything, then slowly move the pen across the paper with out making a mark. You will need to find a way to prop your palm, wrist or even fingers against something on or near the work piece. Heat can be a major deterrent to holding position. Amazon does sell a product called a TIG Finger, a fiberglass sleeve that you put on one of your fingers over your TIG welding glove to insulate you from the heat. Another item you can use is a brick or two to prop your self against. Being essentially stone, they will not conduct the heat to your hand like metal will.

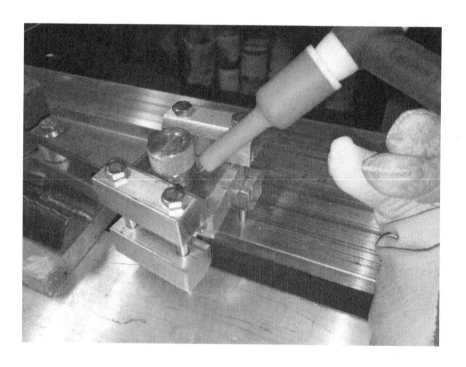

Some times you need to get creative with your hand position. Almost anything is up for grabs, as long as it's safe. Here, I'm finishing up some legs on an aluminum melting foundry steel frame. A large welding magnet provided the proper hand position to make this pass. The long ceramic cups that came in my gas lens kit also are great for getting into tight places, where bulky things like hands would block access.

Foot Pedal Position: Like when driving a car, you want to have a very high-level of control over the foot pedal. It works just like the accelerator pedal in a car. My personal preference, actually, is what came with my Square Wave 200: a full foot pedal where you can position your ankle just slightly behind the pivot point of the pedal allowing very fine control. Other foot Pedals don't allow you to put your full foot on the pedal and can be more challenging to operate. It's not impossible; it's just not as intuitive. Remember, on a foot pedal machine, the arc will not start until you press it down. Note: slight slip on the petal can light off an arc before you are ready or stop one quickly at the wrong time.

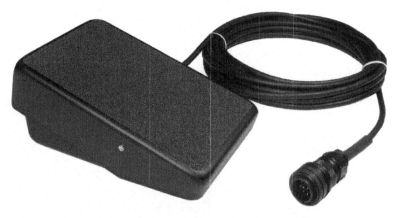

Typical foot pedal operation starts with full pedal to get the arc started, then once established, you may need to let up a bit to keep from over heating the part. At the end you would slowly let up off the pedal to stop the arc.

Gas

Shielding gas: Mentioned Earlier the main shielding gas for TIG is argon. This will suit your needs for steel, stainless and aluminum. Other custom blends of gas are available; however, they are usually tied to a specific material. Helium shielding gas (the 2nd lightest element) will be way lighter than air and rise up away from the weld, letting the oxygen in. (Imagine the speed that a fresh/released helium balloon rises up, and you will get the idea.) Helium is of good use for over head welding, however, most welding done for applications in this book are on the bench or, at most, eye-level. Argon is slightly heavier than the nominal atmosphere and will stay down on the work part. There are usually local welding supply stores like Praxair who carry gas for all kinds of welding needs. Typically, they will "sell" you an argon cylinder kind of like a core charge setup. The cylinder is filled with gas and used. When empty you go back to the same store and exchange it for a full one, only paying for the gas. Unless you're the Hulk, a 110-cubic foot cylinder is about the most you'll want to handle. They can be very difficult to grab and hold on to.

Note: most of these welding supply stores close before 5:00 PM and are typically not open on weekends, so if you have a nine-to-five you will need to adjust your schedule accordingly. Argon is slightly heavier than the normal atmosphere of nitrogen and oxygen and works well for items on a bench where the shielding gas descends on the part and settles protecting the part during welding.

For additional coverage on conventional welds you can use aluminum foil to "tent" or provide additional capture and shielding of the area.

As mentioned in the safety section, the gas cylinder should be secured to your welding machine, cart or to a column in your shop.

Gas hookup

Many TIG machines come with a Gas pressure and flow gauge. If not a welding supply store where you purchase your shielding gas should have some for sale, Amazon and other online options are available as well.

Everything in the gas hookup is a fluid connection so it's a case of tightening connections "just right" **not** wrenching on it like lug nuts.

Anything with an NPT fitting should have Teflon tape on it (install in the direction where it will not get peeled back). Other connections like the one off most gas bottles is a conical type where no tape is used. Everything should be just tight enough to not hear any hissing when the bottle is opened. Note on TIG gas connections everything is typically conventional right-hand threads.

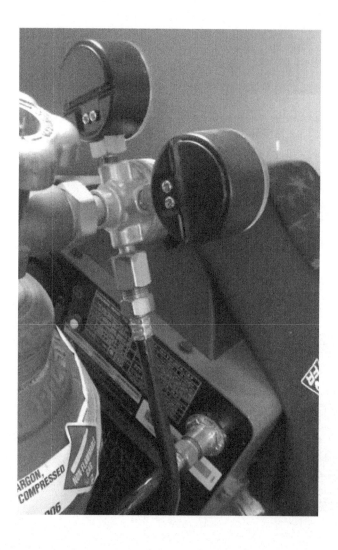

Gas lens: Another edge you can get for producing higher quality welds is to use a gas lens. These typically are not sold with your basic TIG machine but are bought after the fact. While I have a conventional Lincoln #17 torch, I found a generic gas lens kit that works perfect with it. Your stock torch gas flow is quite unstable coming out of the nozzle, as it usually egresses from 2-4 holes straight into the ceramic cup. This unstable flow can lead to uneven argon coverage, and, at higher gas flows it can disturb the weld puddle. A main feature of a gas lens is a 360° filter screen that goes between the ID body and the OD ceramic cup. This filter screen stabilizes the argon flow, turning it laminar inside the cup making for more even coverage and slightly lower argon consumption. The other benefit in ordering an aftermarket gas lens kit is that they usually have several different lengths of ceramic cups. One I find particularly beneficial is the long skinny "ant eater" style that allows you to get into tight corners.

Welding accessories like this Gas Lens kit can also be found on the companion web page www.gouldaero.com/weld

Gas flow: TIG requires a shielding gas to cover the molten and solidifying weld puddle to keep the normal atmosphere particularly Oxygen, from corrupting the weld. The gas flow is a dialed-in balance. To little argon and you will see sparks, even though the welding helmet. Too much and it will distort the weld puddle, and if at too steep of an angle pull in the surrounding atmosphere. The other variable is how thick of material is/how high the amps are. As your amps go up so must your argon flow. A good starting value for .035 - .060" (typical Aero / Auto thicknesses) would be 10-15 cubic feet per hour. Adjust as needed for best results.

Pre and Post Flow: when setting up your first set of welds for the day or after a break the argon line has been filled with the normal atmosphere all the way back to the control valve in the machine. Note: everything down stream from this valve is open to the atmosphere, so after an hour or two the argon in the line will dissipate into the outside atmosphere. In the normal foot pedal TIG setup, you will usually need to "tap" the pedal two to three times to re fill the line with argon. You should always have some scrap available to do a quick test weld to make sure the argon, polarity and current settings are all up to spec and running correctly.

Post-flow is another item to be aware of. The simpler setting machines typically have a post flow timer that's set by the factory. My Lincoln machine is factory set to around eight seconds of post-flow. Machines with advanced settings on the face plate usually have a user lever adjustability of the post flow. The adjustability of post-flow seems to be more of a brand vs. brand deal than a differing price point across all brands. For example, Everlast dedicated TIG machines with adjustable post flow are at a price point half of what a Lincoln machine would be and 1/3 of what a Miller would be at. However, some of these may offer pulse adjustment but may not have AC welding so... no aluminum. After stopping the weld it's important to keep the argon flowing over the hot section of the weld until the post-flow stops. After a few seconds, the weld puddle should solidify under coverage, and you should be all set. You can listen for the post flow hissing as an audible clue as to when it's done.

Setup

Machine and controls:

In this section I will use my Lincoln Electric Square Wave 200 machine as an example, it's a good Average of what you will find with most modern dedicated AC/DC TIG machines.

You should go through the user manual that comes with your machine for high level user functions. In case you lose your manual, bought a used machine without one or would just like an electronic version you can usually find it by a google search typing in your machine's manufacturer name + model number + .pdf. For example, "Lincoln Electric square wave 200 .pdf". Looking out for the manufacturer web site & .pdf in the search results.

Google   lincoln square wave 200 .pdf   Q

          All   Shopping   Videos   Images   News   More   Settings   Tools

[PDF] im10296 square wave tig 200 - Lincoln Electric
https://www.lincolnelectric.com/assets/.../lincoln3/imt10296.pdf ▾ Translate this page
Square Wave® TIG 200. Register your machine: www.lincolnelectric.com/registration. Authorized Service and Distributor Locator: www.lincolnelectric.com/ ...

Square Wave® TIG 200

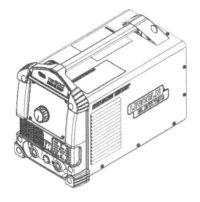

For use with machines having Code Numbers:
12475

Power cord: most modern inverter machines (160-200-amp range) in the USA (60hz) will have a 110v plug for a normal 20-amp outlet and a NEMA 6-50 plug for 220v / 50 amp. These may be two separate cords or just an adapter that goes on the end. **Power source selection is often automated but always consult your user's manual to confirm before plugging in for the first time**, specifically look for if the power is auto select or if there is something you have to adjust a setting, jumper, button etc. in the case of my machine it's an auto select.

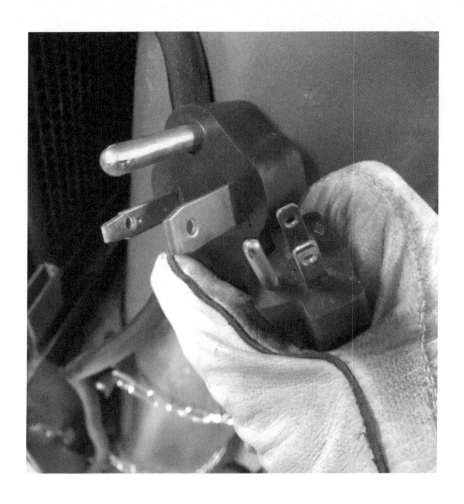

I should note I did see a question on a welding message board about a new welder who was confused on how an AC power source can make DC, setups and so on. As you may imagine the peanut gallery went full force on him. That's not the intent of this book as it's intended to help you the reader get going on the right track.

Some high-level basics on the 2 different power sources and AC/DC selection. 1st welding from a 110v / 20A plug will limit you to around 1/8" material on steels and 3/32" on aluminum.

Going to the 220v / 50A gives you more power so you can weld thicker material so 1/4" steel and 3/16" aluminum. 2nd as for AC vs DC selection that's all in the machine and handled through a selection on the controls.

Torch, ground clamp and foot pedal connections.

Most modern machines use "Dinse" connections for the torch and ground clamp. The more advanced setups have the gas flow through the center of the torch Dinse connector.

Dinse connectors are a quarter turn type. The notch normally points up to plug it in and you give it a 1/4 turn clockwise to lock it in.

With this style the shielding gas flows through the center of the Dinse connector.

Other machines have the gas as a separate connection from the torch.

Foot pedal or torch control is typically a multi pin / thread in to lock type.

Turning on the machine.

Once your gas connections, torch, clamp, pedal and electrical are all plugged in you are almost ready to go. Make sure the ground clamp is on the workpiece or on the table opposite to where you will be welding. Torch should be on a torch stand or stowed on your machine/cart. Foot pedal should be in position where you can use it when you're ready to weld. On my machine the power switch is on the front. Other models may have the power switch on the back near the electrical plug in.

Machine control navigation

Depending on the machine the controls can be very simple or very complex and involved. The simpler controls like the one on my SW200 machine are easy to use while welding, changing setting on the fly. They are typically self-explanatory and don't need to dig into a manual.

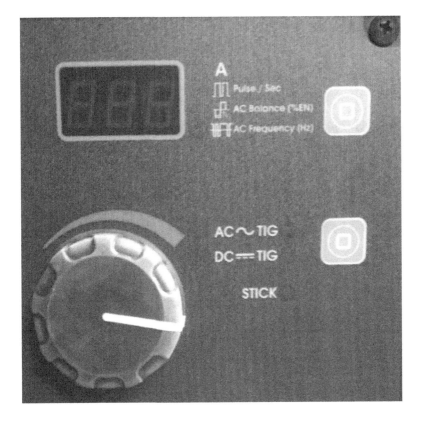

When welding simplified controls can be your friend. Most in process adjustments are usually to amps or pulse.

Others like an Alpha AHP has all kinds of buttons and nobs. They are labeled however you may need to look it over a few times to know where everything is.

And the final type that's gaining more popularity is multi-function screens similar to a cell phone or computer. These take the most studying and a dive into the user manual will be advised.

Whatever control type, you will need to get it onto the TIG setting along with the current type.

Setting the current for the first time.

Here is an example of setting up to weld .035" 4130 chromoly tubing. With no pulse set the amps to 35.

With pulse you usually need to up the amps by 30-40% so with the pulse feature on I would turn it to 50.

Going into the pulse settings anything below 10 pulses per second "pps" or hz usually will give most people a headache. I use 15 hz for most of my work. I should note that my machine maxes out at 20 hz. Some machines like Everlast go much higher than this. Note normal television is at 30 hz and some movies are at 24 hz so going above these values will be difficult to even detect the pulse visually. This is mostly a comfort issue however you should note the higher the pulse frequency the more you will want to dial back the amps as it's not getting as much cooling between each pulse so you don't want to overheat.

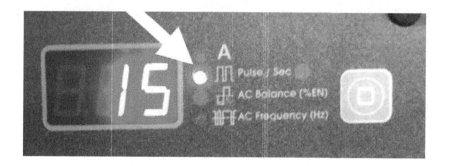

AC setup for aluminum

For this switch the machine to AC mode, all plugs stay where they are.

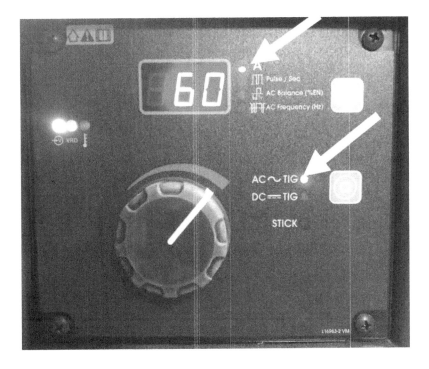

AC function can also have a pulse feature however I have not gotten the benefit out of it as I do with thin wall DC steel welding. The other feature is the wave form offset & frequency.

Wave form offset or balance (%EN) controls the cleaning action, a good starting point is 75%

For tuning it in from this point if the weld puddle is looking contaminated then you need to increase the cleaning action by brings the balance number lower say from 75% to 65% and check for a difference. On the opposite side too much cleaning action will ball up the end of your electrode. The fix for this is to reduce the cleaning action i.e. go from 75% to say 80%.

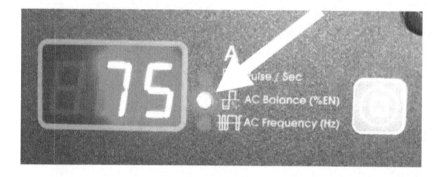

Increasing the AC frequency will create a more focused arc with a narrower cleaning area and weld bead.

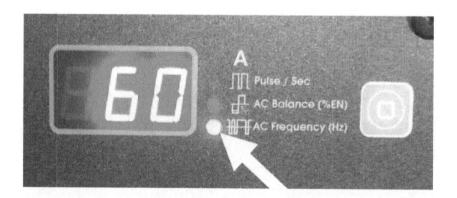

Pulse: As mentioned above pulse can give you more "time" margin welding thin wall till you have a blow out, particularly on a pointed edge on a tubing cluster. This function is another one of the "hacks" that give you an advantage that was not available in entry-level machines a decade or two ago. Another benefit to pulse is you can automate the "stack of dimes" using a "lay wire" technique. In this setup, the filler rod is not dabbed in but set just in front of the weld puddle and the pulse will cyclically enlarge the puddle, grab some filler, and shrink back. Pulse settings around 6 - 12 pulses per second "PPS" can be quite annoying on the eyes, 15 PPS seems to work well for me. 30 PPS would be the same as watching TV at 30 frames per second where you would not even detect that something is pulsating with your eyes.

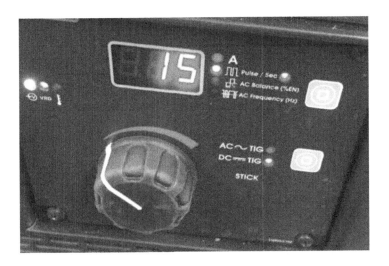

Electrode Selection

There are many electrode chemistries on the market and it can be confusing on what to choose. A brief overview on what to know. Electrode's are color coded to show their alloying chemistry:

Green = Pure Tungsten (Lowest cost however they have high consumption. Usually only used on Aluminum "AC" welding)
Red = Thoriated (1.7 – 2.2% thorium, long lasting and multiple types of metal however it is Radioactive)
Orange = Ceriated (1.8 – 2.2% Cerium similar to Thoriated but not radioactive)
Blue & Gold = Lanthanated (Blue is 1.8 – 2.2% & Gold is 1.3 - 1.7% one of the most universal "go to" types for doing AC & DC work, they are also long-lasting price point is around the same as Red & Orange)
Purple & Grey = Rare earth (these can be more expensive then the other options but along with the Blue band is common at welding supply stores)

So, with this what do I recommend of a reader of this book, TIG welding Steel, Stainless or Aluminum in their Garage? For me the answer is Blue band lanthanted, I want a "go to and go" solution that covers all my needs. For what we do it can do anything the Red band can do with out radiation! While it's a small amount everything adds up given enough time and exposure.

Sizes:

There are 4 common sizes most TIG electrodes are sold in and 3 that are applicable to the 200-amp machines covered in this book.

.040" (1.00mm)
1/16" (1.58mm)
3/32" (2.38mm)
1/8" (3.17mm)

The last (1/8") is typically beyond the usable range for 200-amp machines.

A safe rule of thumb is the electrode diameter can approximately handle up to 2X the amps, for example:

.040" "40" = 80 amps max
1/16" "62" = 120 amps max
3/32" "93" = 180 amps max
1/8" "125" = 250 amps max

If the electrode is dissolving in front of you quickly its too many amps for that size and you will need to go up a size or decrease amps.

Next thing to know for a given material, thickness and TIG machine settings going up a size in electrode will put more heat into the weld for a given amount of time, going down a size will put less heat into a weld.

Most of my work is thin wall .035" 4130 Chromoly tubing and .040" Blue Band Lanthanated is my go-to. With 50-55 Amps & 15 PPS Pulse settings I have good margin in "time" before I have a blowout. Going up a size to 1/16" with everything being the same greatly reduces this margin due to the added heat.

TIG torch total length & weight:

I have found you can get much more control and better access using the smallest, shortest, lightest, torch setup you can get. On aircraft projects I usually like to use "half length" electrodes that conveniently fit into a #9 short torch with a gas lens.
You can split electrodes at home but there is a right way, a wrong way and an easy way.

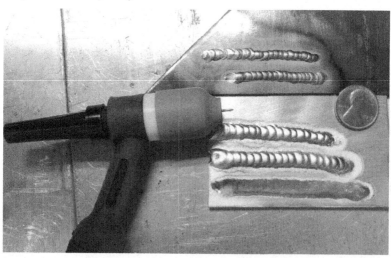

1st the right way:

You will need to use a carbon cut off wheel or the edge of your grinder i.e. a circular cutting motion that "files away" the material till its in to parts. I can be a little time consuming but it assures you don't put any cracks into the electrodes. While the electrodes look like steel while they are actually quite brittle and internal cracks will mess up your arc, wasting a lot of time looking for "why did the arc get all messed up tip looks fine". Remember once split, make sure to do a proper tungsten grind job with the grooves going in the same direction as the length of the electrode.

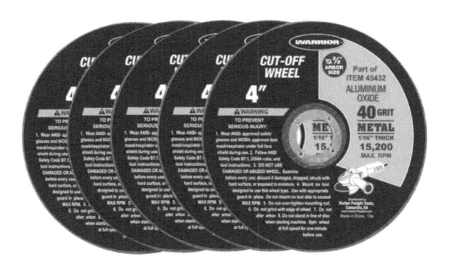

2nd the wrong way:

Trying to split it in half over a sharp edge, using dikes etc. This puts internal cracks in the electrode as mentioned above.

3rd the easy way:

Buy them already cut short, while most electrodes are sold in 10 packs in the 6-7" length there is one company that sells half-length ~3" blue band electrodes in a high count called JSP ARC. They are my go-to for half-length and full-length electrodes, they even offer combos.

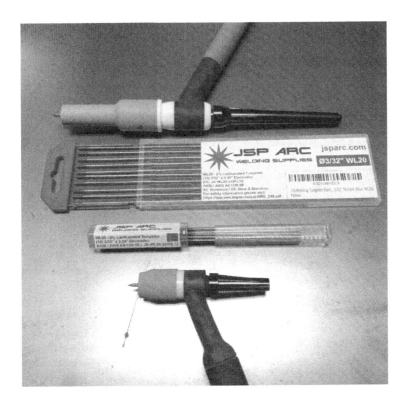

Electrode shaping: While there are a few ultra high cost products for shaping electrodes on the market the key thing to have is a dedicated machine to sharpen your tungsten. While not the most compact, a bench grinder is usually the lowest costing way to go. Sharing a grinder for use with other grinding needs can contaminate the electrode, making for poor arc quality. Chucking the electrode in a cordless drill makes for a more even profile around the wheel. For your own safety and ease off shaping you should always point the electrode in the same direction of grinder turn NOT into it, you don't want to have it catch and make a projectile. Respirator and eye-protection should be worn at all times when sharpening electrodes. Older TIG electrodes were alloyed with thorium (red band), a radioactive element, however the tungsten it's self is a heavy metal and is present as the prime alloying element in all TIG electrodes. The point to the end of the taper should be approximately 2 electrode diameters back.

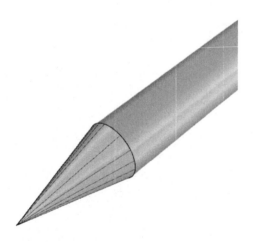

On steel you can leave the point sharp for Aluminum welding you will need to blunt the end approximately 1/4 - 1/3 electrode diameters off the tip.

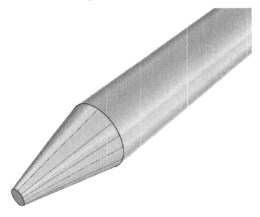

The serration pattern from the grinder should run parallel to the long axis of the electrode, not cross wise. This has an effect on how stable and controlled your arc will be coming off the end of the electrode. Remember electricity, flows on the outside of a wire or electrode not through it so you don't want to make any speed bumps for the electrons to go over.

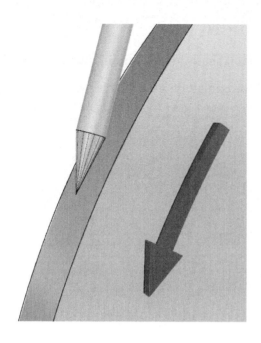

Jigs: When it comes to part alignment and the ability to control heat, jigs are the most important factor. When it comes to welding steel, Aluminum jigs work quite well. Aluminum has a much higher heat conductive capacity over steel and helps soak up heat spikes that could blow through thin wall steel. It also acts like a heat reservoir, allowing for an easier time welding with fewer amps over the long run. I also use two 3' x 6" x 1/2" aluminum plates that sit on my weld table that are the main surface I weld against. For items like the welded side of an airplane fuselage or other simple setups non-treated plywood and wood blocks can be used to jig up 4130 tube structures. Make sure to pay attention to any smoldering or fires that could arise. Keep the jig and part out in the open and away from any combustibles.

Weld Table: There can be some very high-end weld tables out there like the Certiflat series from https://weldtables.com These have a lot of useful features, and are great for pro shops, however, they can be quite expensive purchase and ship. Also since they are a kit, they are project, not a product. If you have the time, money and space it's one of the best way's to go.

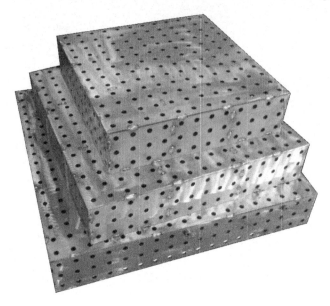

For someone working with small parts or just getting started I would suggest the Member's Mark Work Table with Stainless Steel Top from Sam's Club (#44648), (also available from Amazon.) I've had great luck with it. It's fairly light weight, easy to assemble and costs around $120. There is also a wire mesh shelf underneath where I store my metal cutting chop box and welding helmet in a clear plastic tub. It comes with wheels so it can be rolled around the shop.

Some "mystery meat" aluminum plates can also be very useful as part of your weld table. The ones seen below came from the useable drops area at Alro metals. They were not marked, but for this, they'll work fine.

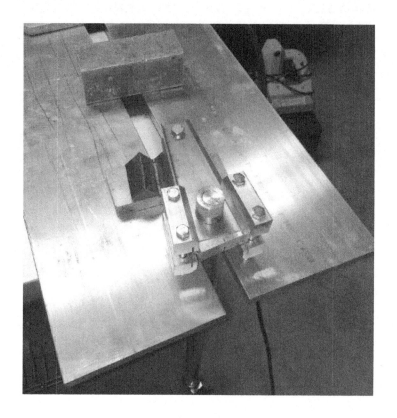

Welding magnets can be very useful with steel welding. Harbor Freight and Tractor Supply have a good selection. They can be used on both tubing and flat components and even as a way to get a better hand position with a torch or filler rod hand.

These magnets should only be used to hold parts together for tacking. The heat from full part welding can melt the magnet media between the two metal plates.

Additional jigs and fixtures: With 4130 and other tubular structures, it's common to build what you can, flat, then turn it to 3D. Here you can see turn buckles being used in conjunction with wooden blocks to hold this aircraft fuselage truss in position as it gets welded.

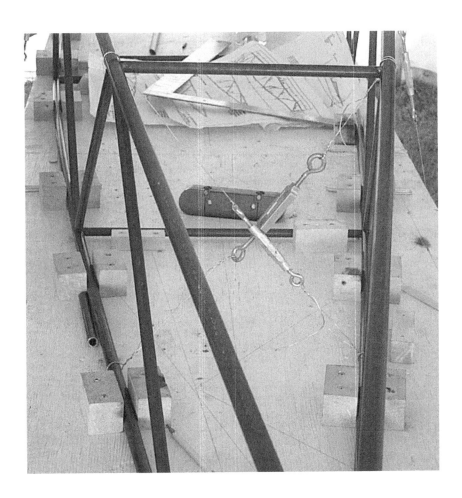

Protective Equipment

The companion site www.gouldaero.com/weld has links to buy some of the protective equipment listed in this section.

Helmet: As mentioned earlier, this is another area not to be cheap with. Most modern helmets have auto darkening that lets you see your setup and almost instantly goes into full shade when the sensors in the lens detect the arc from the electrode to work piece. Some of these helmets have adjustable shade levels. If you can still see the bright light when you close your eyes your shade is too light and you need to adjust it to be darker. These helmets are either solar powered or a combo of solar and button battery powered. Since my shop can get quite dusty and my Lincoln Viking helmet is not cheap, I keep it in a dedicated clear plastic tub. This can also help charge up the solar before use.

Note When shopping for a helmet, make sure you can adjust the darkness of the shade. Low amp/thin wall will require a lighter shade setting compared to high amp/thick wall work. It should be light enough to see what you need to see, but not too light where you can still see the arc when you close your eyes (like what happens when you stare at a bright light.)

Cheater lens: Clear visibility of the weld puddle is everything to making high-quality welds. While my vision is average, I wanted a way to optically zoom in, kind of like what a jeweler does. My first stop was to Wal-Mart to check out reading glasses. I tried all kinds of magnifications but found the clear focal length to be way too close to my face to be practical not to mention heat the and fumes so that was a no-go. Next, I try was jeweler magnifying glasses from Amazon, Like with all online purchases, the best you can do is check out the pictures and, read the description and reviews. It turns out these had an even closer focal length than the reading glasses. They did come in handy when putting in watch batteries but they weren't good for welding. Success came with a trip to the local Tractor Supply where they actually sell a good amount of MIG and Stick welding equipment. I found a Hobart "cheater" lens that I was able to take out of the packaging to check its focal length. It turned out to be perfect. I though would have to rig up some way to attach it to my Lincoln Viking series helmet turns out it has a tab and slot design in the helmet where the Hobart cheater lens just pops in. So, even on helmets there is some compatibility across brands.

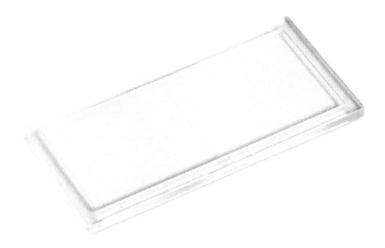

Fume extraction: Professional/purpose-built fume extractors are made by big players in the market like Lincoln, Miller and some specialized companies. They are typically, a large box on wheels with a two segment or more pickup arm and a funnel nozzle on the end so it can be placed near the active weld to suck up the fumes. In summary, they are a fancy vacuum cleaner and filter quite effective and turn-key, however, they usually cost more than a welding machine.

You can also make your own fume extractor with a 20" x 20" box fan, a Charcoal 20" x 20" air filter duct taped onto the output side of the fan, and an adapter plate that covers up the free corners on the input side of the fan so the fumes don't escape out that side and are forced to go to the filter and get trapped. This setup should be used in combination with an open window and box fan or some other method of shop ventilation.

Respirators: I bought a Miller welding respirator at the local welding supply store. While it is quite compact, it still sticks out a good inch more than my nose. This pushes the helmet to a slightly open position, letting in arc flash under the helmet, not to mention, more fumes. This option may work for some people (likely those with smaller heads, bigger helmets, more adjustability, etc.) but not for me. I ended up using fume extracting fan and filter setup.

Welding Jackets and Welding Sleeves: As mentioned earlier, protecting exposed skin against electrical contact and ultra-high-intensity light exposure is critical to your safety. I ended up getting a welding jacket from Praxair. It's a lime green color, low cost, but it does the job. Living in South Florida, it can get warm wearing it, but it beats the alternative. I recently made the move to welding sleeves for most of my welding work and they are another alternative to a full welding jacket that can keep you cooler, all while having a heavier material where its needed. All clothing you wear should ether be flame retardant or flame neutral. Stick to cotton, denim, and leather; avoid polyester and other synthetic materials. Your attire can be as simple as the Praxair "green coat special" all the way to top brand items. Either one works as long as your skin is protected from the UV light and heat. It's hard to be the "cool" dude that does not wear protective clothing with a 2nd degree sunburn all over. Also darker colors tend to reduce the amount of light that comes from up and under your welding helmet. I tend to wear a heavy black or dark blue cotton shirt that fits close to the neck.

Filler Rod

Each type of metal is going to need a specific type of filler rod. The common filler rods can be purchased from Amazon, eBay, Aircraft Spruce and other online sites, along with local welding supply shops like Praxair. The filler rod typically comes in the following sizes 3/32", .045" 1/16", 1/8", all 3' long and is sold in packages of one, five and ten. Hopefully, your filler rod stock comes in a plastic tube. If it just comes rolled up in paper or in a cardboard tube you can buy filler rod storage tubes at most welding supply stores and on-line.

Materials

Since this book is focused on the beginner/home DIY welder, we should note that there are some common materials for welding projects, some metals that should not be welded, and some ultra-advanced stuff that should be saved for another book.

4130N Chromalloy is one of the most common structural steels for aircraft fuselages, engine mounts, roll cages, motorcycles and bikes. 4130 is easily TIG weld-able and is one of the best materials to learn on. There are 3 main sources for 4130: USA, Germany and China. If it's something critical like an airplane fuselage you may want to stick to USA or German material. To take it up a notch, you can even get a material cert's report for an up charge. Welded parts on certified aircraft have to get this report with each batch of material. Industry consensus is 4130N under 1/8" wall thickness does not need pre-heat or a post heat treatment. This is a major plus, as airframe construction is typically in the .035" - .062" range, with auto/motor cycle's being slightly heavier wall for most of the structure.

ER70S2 is the prime filler rod of choice for 4130.

6061-0 and T6 Aluminum is the common choice for "structural aluminum" The number on the end denotes the heat treat; "0" is soft condition material. Whenever you arc weld metal, it takes it back to near cast properties in the area. In the case of welding 6061-T6, the heat affected zone and the weld would be around the 0 condition for yield/ultimate etc. Specific heat treatments can raise material properties, however warpage must be carefully checked and fixtured against or corrected.

3003 is a commonly available low alloy Aluminum.

5005 is an improvement over 3003, often used to make aluminum fuel tanks.

4043 is the prime filler rod of choice for 6, 5 and 3 series aluminum's

300 and 400 stainless steels: With in these two families of stainless steel, many are readily weld able using TIG. The 300 series is generally lower cost and lower strength compared to the 400 series. Filler rods are offered in a ER3xx or ER4xx format. Try to best match or match the material to the filler rod on stainless.

Raw Material Supply

Depending on the job and if you are just practicing, there are many suppliers of steel and aluminum available.

For 4130 N tubing the two best sources I've found are:

Aircraft Spruce (ACS). They do internet and mail order sales of raw materials and components for airplanes. They also have locations in Georgia, California and Canada.
http://www.aircraftspruce.com/

Wicks Aircraft. Similar to Aircraft Spruce (like Lowes vs. Home Depot) They are located in Illinois.
http://www.wicksaircraft.com/index.html

Get on their mailing list and sometimes they offer sales and reduced or free shipping etc.

If you happen to make it to Oshkosh Airventure or Sun 'n Fun, you can place an order at the show, and usually they offer free shipping.

ACS and Wicks also offer 6061, 3, and 5 thousand series aluminums.

For general metal supply one of my Favorite suppliers is Alro
http://www.alro.com/. They supply many of the CNC shops along
the East Coast and have locations in several states. They also have
what's called "usable drops" = cut offs from major orders. For
example steel box tubing stock normally comes in 20' lengths, a
customer may order 16' feet of it, leaving the remaining 4' back in
the usable drops area. They have customer shopping rooms where
you can go through the organized materials and pick out what you
need based on general shape/size/material etc. Many of these
materials still have their markings on them, particularly with 6061-
T6, due to usage rates. It's a very good location to buy square and
rectangular tubing to build tooling fixtures, trailers, specialty stuff.

If you need something really specialized for raw material supply, there is always McMaster Carr, however, some of this stuff can get quite spendy with them. https://www.mcmaster.com/

Learning the Craft

While professional TIG Welders make it look like almost an effortless juggling exercise, it's really a set of individual specific skills all being used at the same time. Here is what you need to learn in order to become successful:

1) Have a good amount of scrap steel acquired (cleaned and prepped), along with the working machine, all your gear, a full argon cylinder and so on. Steel is more forgiving to start learning on compared to aluminum and other metals.

2) The TIG Torch needs to be connected to the DC negative "-" port and the ground clamp to the DC Positive "+" port. Some TIG machines like the Lincoln square wave the gas comes through the center of the electrode port on the box so there is only one thing to plug in on the torch side. Other machines may have an electrical connection and a separate gas connection.

3) The electrode should be ~2:1 taper i.e. a 1/16" electrode the cone would be 1/8" long. (see electrode sharpening section for more details.)

4) Depending on your cup configuration, your electrode should stick out ~ ¼" – 3/8" all connections should be hand tight and <u>do not</u> use any mechanical leverage to tighten anything on torch side or electrical plugins on the box.

5) For a given thickness of steel, you will need a given number of amps. It's best to start without pulse on. With this setup, you will need one amp per .001" of metal thickness on steel and 1.2 – 1.5 amps per .001" of aluminum. So, if your welding .035" 4130 it would take 35 amps, .062 would be 62 amps. Note: if you're on an inverter machine plugged into 110V, you will likely be limited to 1/8" material. This is usually the max that you would see on typical 4130 chromalloy structures for most aero and automotive applications.

6) The grounding clamp should be clamped direct to the work piece (what you intend to weld) or on the welding table

either way, opposite from where you are. (See the safety section at the beginning if you need more details)

7) The machine should now be turned on along with the argon gas valve on top of the cylinder. When you press the pedal, you should hear a hiss of gas coming out of the torch. You should press the pedal at least two times to purge the atmosphere out the lines.

8) The first type of welds you should start with are just on a single piece of steel. .062 - .125" thickness is the best to learn on.

9) Arc initiation: with your foot NOT on the petal, the TIG torch should be ~ 1/8" off the work piece with your hand resting against something. Torch angle should be perpendicular to 15° of tilt to allow you better visibility of the electrode tip. With the helmet down, you can now press the pedal and the arc should light off. This first baby step should be for only a few seconds. Your helmet should auto darken instantly, and you should not have any eye discomfort as the helmet is intended to take out the intense light but still let you see the puddle and some of the surrounding area. To stop the arc the correct way slowly let off the pedal. Pulling the torch away from the work piece also will work. You will lose weld quality with this method, but it works in a pinch if you're trying to learn all this stuff at once.

10) Repeat the previous step on fresh areas of metal getting used to turning the arc on and off. You should not get any

burnt-out spots if only in position for a few seconds and should not see any major sparks flying off the weld. If you do, you may need to dial up or down the gas flow.

11) Once you get used to lighting off and shutting down an arc the next step is to move the arc along the work piece. Direction of travel should almost always be going towards your other hand. You will need to maintain proper hand support through this process. The end goal is to have an even bead of weld on the single sheet of metal.

12) The next step is to repeat the previous step but start adding in filler rod with the other hand. The Filler rod should be similar or smaller in size to the thickness of the base metal your welding. You should start with the dab method and closely notice the effects on the puddle shape and spread as you add filler rod. When you add filler, this temporarily cools the puddle and reduces the puddle size.

13) Once you get comfortable with Arc control, moving the arc and adding filler, it's time to start to join two pieces of metal. A but joint is the easiest type to start out with. The fit should be 0 - .040" wide. Using all the steps and skills you have learned so far you can create a puddle that bridges the gap between the two metal parts. Tack together both ends before performing the run.

14) The next joint to go to is a lap joint, then try tubing joints. The more complex the geometry, the more repositions, starts and stops you will have to do.

15) Practice with the torch in both your dominant and non-dominant hand, along with multiple positions and angles. Professional TIG welders earn their keep by the adaptability they have to new setups and, angles their ability to get into and hold and difficult positions to get the job done.

16) Practice makes perfect. I went through two dozen test parts before I did my first "real" airplane part with the same type of joint.

TIG Welding Methods

Where to "point" the heat: If both pieces of metal are the same thickness and are not supported by a jig, then you would want to have your electrode oriented half way between the two parts. However, if one side is thicker than the other or has jigs or heat soak blocks against it you will want to have the heat from the arc concentrated on this part. It's all about balancing the heat in both parts. On thin wall materials a massive, imbalance of heat can lead to a blow out, even with pulse.

Tacking: with any kind of welding, it is critical to tack the extreme features of what you are welding in order to ensure proper alignment and reduce warpage. Tacking patterns are similar to tightening down a wheel on a car. The goal is to jump from "opposite corners," back and forth, until all details are tacked in. One of the benefits of TIG is you can start and stop a weld any time with out re-prepping the surface. If your part is sensitive to warp-age then you can keep processing short welds .5 - 1" following the same pattern used for the tacking setup until everything is complete.

Tacking patterns should be evenly distributed like torqueing a wheel to control distortion and movement. Below is a tack welding sequence for a 6-bar frame.

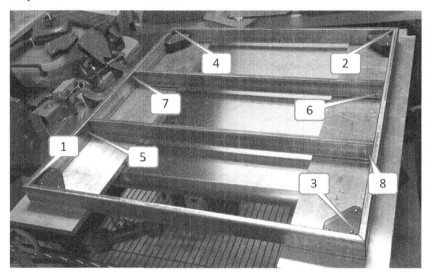

Puddle Start: Typically, one of the sides of the weld will start to liquefy into a puddle before the other. Once this happens you need to move the torch more towards the other side. This is a balancing act. High amp setting will make the puddle form faster, but things can run away quickly; lower amp settings will take longer to form the puddle, and will have longer "puddle ready" times when applying filler rod.

Dab in filler Rod: This is the oldest and most conventional method and works with or with out pulse. Its going to be your go-to method for most transformer machines. Unfortunately, you actively have to be controlling both hands at the same time. With this method the puddle is generated between both parts, and, once you see the puddle start to crown, you add filler rod into the puddle for a second or two, then back the filler rod out from the puddle.

Lay Wire: This method is typically coupled with pulse, walking the cup, or free hand methods. This requires less control over the filler rod hand compared to the dab in filler rod method. More concentration can be put onto the torch hand.

Walking the Cup: is typically done on fillet welds, where the ceramic cup of the TIG torch is nested between the two parts. The cup is walked back and forth between the two sides, generating an "S" pattern. This method can be used with lay wire, as the act of walking the cup provides the puddle control. The variables are mostly left to the torch hand, so, in some ways, it can be simper.

Free Hand: this is similar to walking the cup but with no guides for the cup; its purely the motion of the TIG Torch with out the ceramic cup resting or referencing anything near by. On a tubing cluster you may start off walking the cup in the sharp angle between tubes then transition to free hand once that part of the cluster becomes flatter to maintain the profile and style of the weld.

The Finished Part

Post-welding cleanup and protective coatings/paints: The more practice you get, the more you will want to show off your welds rather than grind off any boo-boos. Some things, like exhaust systems, are left as is, other things like welded fuselages, roll cages, and engine mounts, get painted.

The best preparation method out there for tubular structures is sand blasting. If you have small parts like control components, you can usually get those into a sand blasting cabinet. For larger components you will have to wear a protective suit, and the sand blasting is done in the open. There are commercial services that will sand blast your parts for you if needed. Good alternatives are wire wheel brushes attached to grinders, assuming you don't dig into the parts, and dual action "DA" orbital sanders. The surface will need tooth for the paint to stick and will also need to be free of oils and other contaminates.

Unless your going ultra fancy and custom I've found Rust-Oleum Universal spray paint to be one of the most durable and best finishes you can get from normal box stores. Part of it is the paint chemistry being just very good; the other part is the spray nozzle that produces a "fog" of paint just like a pro spray rig compared to the older and more typical "super soaker" spray output of older products.

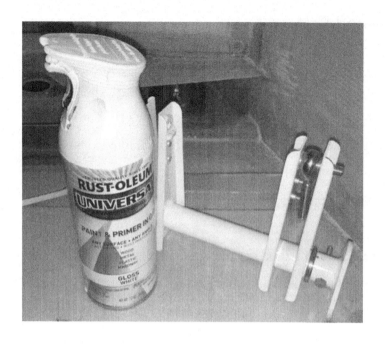

Note: For aircraft, auto, and marine applications, I would not powder coat your welded components. Powder coating is a thermal-cured ceramic, prone to cracking since the coating can be stiffer than the base metal and is typically more brittle than the actual structural part. These cracks in the powder coating can allow corrosion into your part leading to increased maintenance or even part failure.

Zinc Chromate: A mainstay of the aviation industry for many years, however I have found it to be more challenged in grip to the surface compared to more modern paints and coating. This may be ok for something you plan to sand off later but I would look into more modern materials to get a more durable end product.

Epoxy and Polyurethane: An excellent modern choice for large jobs, grip is excellent along with finish durability. You will need a spray rig for this (paint gun, compressor, dryer etc.) It is important to clean up the reusable parts of your paint gun right after finishing the job as they will become junk if the coating is left to dry in the paint gun. The pre-mix and post cleanup is the reason to leave this option to large jobs like a whole car or airplane frame vs. just one simple bracket.

Trouble shooting

Problem: Weld has charring, orange rust, spatter, visible prosody "holes", makes a bunch of noise with DC/steel.

Solution: Shielding gas. Most likely gas flow is set too low. Make sure to purge your line and get a good weld on scrap before going over to critical parts. You should hear the gas hiss when you step on the pedal. No hiss means no supply pressure an open valve. This means the tank is likely empty

Alternate Solution: The area you're in is too windy and is blowing away the shielding gas. If you can't change the location, use wind blocks to reduce wind in the weld area.

Problem: Arc is not concise, "dancing" all over, no control.

Solution: The most likely candidate is that the tungsten has become contaminated with a contact with the weld puddle and needs to be re-grounded.

Alternate Solution: tungsten grinding marks must be parallel to the rod. If they go perpendicular, this can fowl up the arc.

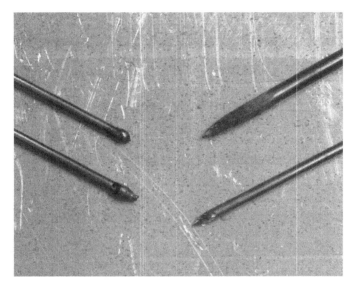

Problem: blowing holes through thin material.

Solution: The amps likely set too high. Also, use the pulse feature for more margin "time" until a blowout occurs.

Problem: Weld is under-cutting parent materials

Solution: More filler is needed = slow torch speed, increase filler rod speed.

Problem: Seeing arc light even after you are done welding/helmet is up/eyes are closed.

Solution: Helmet lens is set on too light of a setting. Lens battery is low or dead.

Problem: Can't see weld puddle

Solution: Try reducing the shade setting on the helmet/lens.

Problem: Weld puddle start is taking a really long time to get going.

Solution: Increase amps and or decrease pulse.

Problem: Your torch or filler rod hand is too shaky or grounding out the electrode too often

Solution: Your hands must be stabilized against something fixed.

Links:

Welding Machines

http://www.lincolnelectric.com/

https://www.millerwelds.com/

http://www.everlastgenerators.com/

http://www.ahpwelds.com/

http://www.esabna.com/us/en/index.cfm

http://www.eastwood.com/welders.html

Welding Supply

http://www.praxair.com/

https://www.amazon.com/

http://www.ebay.com/

Material Supply

http://www.alro.com/

http://www.aircraftspruce.com

http://www.wicksaircraft.com

http://www.onlinemetals.com

http://www.speedymetals.com

Alternate Welding Accessories

http://www.harborfreight.com/

https://www.tractorsupply.com/

For aviation, and automotive related technical content, how to's, reviews, products, and projects please visit **www.gouldaero.com**

And for the dedicated companion web page for this book, please visit **www.gouldaero.com/weld**

About the Author

I grew up in the construction industry and had a very early interest in aviation becoming involved with the Experimental Aircraft Association EAA in 1993. While I picked up aircraft composites and wood working early welding was an elusive skillset for a good amount of time. Through the years Major Aircraft shows like Oshkosh/Airventure and Sun 'n Fun provided good hands on opportunities, along with the assistance of fellow EAA-er William Wynne, helped me get down the safety and 101 basics. The engineering curriculum at Embry-Riddle had very little hands-on fabrication exposure. Going down the white-collar engineering career path did not find too many hands-on opportunities working for aircraft and engine OEM's. The modern engineering organizations also are perceived to frown on engineers learning hands-on manufacturing skills, but are quite appreciative when they already have the skills and could apply them to the job: quite the catch-22.

The pivot came when deciding on whether to sign up for a tradesman community college course in TIG welding that would cost thousands of dollars, have a rigid schedule that work and home life would have to bend around, and have classic academia and tradesman school, where a large portion of the classes are spent on things OTHER THAN mastering the skill of TIG welding, and in the end, have no equipment at home, or, by knowing the safety information up front invest the money in buying a machine and spending the time at home using home study methods to learn the details along the way. When stuck between a rock and a hard place on one of my homebuilt aircraft projects needing some control components welded, being full days drive away from his welding contacts the thought of being the damsel-in-distress seeking a local hireling of the parts to local "un-knowns" was the straw that broke the camels back. Up until this point, my hands-on welding learning path was quite passive. The switch was flliped and TIG welding was a skill that I had to master. This was achieved by binge watching 100's of hours of welding videos online and picking the brains of the far away welding contacts I would see at the annual aviation shows.

This was all months in advance of the Annual Sun 'n Fun air show where I knew Lincoln Electric would have show specials on there product line up. Being all pumped up for the event watching all the videos getting my chance to try out the machines at the show, this year I was prepared. When I finished ship lap pass on 1/8" steel the welding instructor had me do, he said "very impressive. Have you done this before"?

This was the verification that he was on to something. I bought a Square Wave 200 TIG machine at the show and after a month at home was welding .035" 4130 like a pro.

This book is a practical summary of everything I have learned since 1992, along with the many gotchas that can hold you back from being a successful TIG welder in a concentrated, need-to-know format.

Hopefully you learned new things about the TIG welding process, equipment and setups with this book, along with giving you the motivation to get into the skill, art and science of TIG welding.

Spencer Gould

EAA 466275

EAA Technical Counselor

Blog / Web Site: www.gouldaero.com

TIG Book companion Web Page: www.gouldaero.com/weld

For other DIY / home work shop books from this author check out:

Composite Structures & Construction:

Modern Methods in Wet Lay-up & Prepreg Construction

for Aerospace / Automotive / Marine Applications

ASIN = B07466P3SW / ISBN-13: 978-1973890133

Made in the USA
Columbia, SC
13 November 2020